Arithmechicks
Explore More

Ann Marie Stephens

Illustrated by Jia Liu

ASTRA YOUNG READERS

AN IMPRINT OF ASTRA BOOKS FOR YOUNG READERS

New York

10 chicks and 1 mouse
hustle to the hills.
Mama hatched a hiking
trip up Blackberry Ridge.

Best of all, cousins are coming
to keep them company.
The Arithmechicks cherish
family time!

The bus pulls up. Ducklings pour out.
Beaks squeal and wings squeeze.
Together again at last.

While birds bustle about,
the bus zips off then—

"WAAAAH! MY sheep!"

One duckling left her stuffie on her seat.
Her Mama says, "Don't worry.
The bus will be back in a bit."

17 hikers want to head out.
1 wants to wait for the bus.

17 is greater than 1, but she won't budge.

Her siblings try to soothe her.
Her Mama tries to move her.

Mouse makes things better, for now.

Let's go!
There are 9 packs with drinks and snacks.
15 trekkers want to carry them.

9 is less than 15,
which means they'll have to take turns.

Mouse has an idea for a surprise.

Wings swing as feet crunch leaves.
They stop on a bridge to look at a stream.
Water flows. Oh no! Tears do too.

Try a close-up view.

There are 7 big fish and 3 little fish.

7 is greater than 3.
But every fish swims fast!

Mouse picks up pebbles for the surprise.

Wildflowers wiggle in the wind.
Wild wanderers wiggle
and giggle as they walk.
Except for one duckling
with a quivery beak
too busy looking for the bus.

Mouse and birds find fallen petals.
8 pink and 8 yellow,
the same amount of each.

8 is equal to 8.

This is some secret, Mouse!

As the hikers climb higher,
they play games with nature.
They spot ants and acorns.
One Mama asks, "Do you see
any blackberries?"
"Not yet!" all the cousins say.

"Do you see the bus?"
one duckling shouts.
"Not yet," they repeat.

The Mamas are out of breath
and need a break.

Chicks want to make music while they wait.
There are 5 ducklings and 10 chicks.

Though 5 is less than 10,
chicks let the ducklings choose first.
They play drums and chicks sing.
Mouse is the maestro.

Sticks drop to the ground.
Eyes do too.
One duckling feels less than nice.

Maybe talking would help.
So they listen to tales about Sheep.

Mouse gets an extra special idea.

The Mamas revive the hike.
"Where are those berry bushes?"
One Mama thought
they were by the rocks.
The other Mama thought
they were on the slope.
Eyes look up and out.

"There they are!"
One cousin finally
notices nature.

Plip, plop, berries drop.
Everyone stops to count.
There are only 14 berries.
Keep picking to fill both baskets!

14 is less than 18 hungry bellies.

Mouse seizes more pieces for the secret.
What are you up to, Mouse?

Cousins munch berries
and make more memories.
How sweet!
Now for one more treat.

Chicks and Mouse hunch over
and hide their surprise until . . .

TA-DA!
One gift for every duckling and for their Mama.

6 is equal to 6.

Mouse made a special sheep to keep.

The Mamas clap. That's a wrap.
Time to head back.

Wait—

The ducklings have a surprise too!
They brought brownie bars.
12 bars in all but chicks want to share.

Since 18 is greater than 12,
bars become bites.

18 > 12

Hiking down the hill takes less time with one duckling in charge.

The bus is back. Cousins cuddle cousins. Ducklings board with berries and buddies. Wings wave. Some beaks quiver.

Mama says goodbyes are good
because they lead to more hellos.
And that's the greatest.

The Arithmechicks and their duckling cousins can't stop comparing numbers, especially when their Mamas are giving out hugs. Each chick, duckling, and Mouse wants an equal amount.

Arithmechicks (and kids) use greater than, less than, and equal to, when comparing numbers. There are also signs to place in between numbers like this: 2 < 8, showing that 2 is less than 8. You probably compare numbers more than you realize. When you are given a choice between two cookies, do you choose the one with more chocolate chips or less? That depends on how much you like chocolate! If you and a friend want to share pinecones you collected, you would split the amount into two equal piles. Whether you are in nature or at home, you can always have fun comparing numbers.

Less than (<) is used when the first of two numbers or group of objects has a smaller value than the other.

7 < 19

Equal to (=) is used when both numbers or groups of objects have the same value.

12 = 12

Greater than (>) is used when the first of two numbers or group of objects has a larger value than the other.

8 > 4

For all my former students: Go exploring! —AMS

For the explorers out there, who never hesitate to find an adventure. —JL

Astra Young Readers • An imprint of Astra Books for Young Readers, a division of Astra Publishing House • astrapublishinghouse.com • Printed in China • ISBN: 978-1-63592-599-9 (hc) • ISBN: 978-1-63592-600-2 (eBook) • Library of Congress Control Number: 2021925706 • First edition • 10 9 8 7 6 5 4 3 2 1
Design by Barbara Grzeslo • The text is set in Futura medium. • The illustrations are done digitally, and the artist made textures by hand to apply in the digital illustration.